Rainmaking

With the Help of Yajña

By

Dr. Ravi Prakash Arya

Amazon Books, USA

In association with

Indian Foundation for Vedic Science

1051, Sector-1, Rohtak, Haryana, India, Pin -124001

Contact No. 09313033917; 09650183260

Email: vedicscience@rediffmail.com

vedicscience@hotmail.com; vedicscience@gmail.com

Fifth Edition
Kali era: 5119 (c. 2017)
Kalpa era : 1,97,29,49,119
Brahma era: 15,55,21,97,29,49,119

ISBN 9788187710370

Rainmaking: A Historical Perspective

Rainmaking is not an idea of recent origin that has downed upon the scholars of the twentieth century only. For long people have been attempting to induce rain by a variety of methods such as the rain dance, singing of rain songs, offering prayers, the sacrifice of men or animals lighting of fires, firing of cannon and the production of electric discharges by kites. But none of these methods could stand the test of time and proved to be worthless exercises and so by and by receded to the background and went out of vogue. All of those methods were based on the false belief and assumptions of the people and had no concrete or scientific basis behind them.

Modern Method of Rainmaking

The methods of modern rainmaking are, however, said to be based on a knowledge of the physical processes of rain formation. Modern rainmaking experiments are based on three main assumptions,

1. That either the presence of ice crystals in a supercooled cloud is necessary to release snow and rain by the Wagener-Bergeron process, or the presence of comparatively large water droplets is necessary to initiate the coalescence process.

2. That some clouds precipitate inefficiently or not all because these agents are naturally deficient.

3. That deficiency can be remedied by seeding clouds artificially with either solid carbon dioxide (dry ice) or silver iodide to produce ice crystals, or by introducing water droplets or large hygroscopic nuclei.[1]

Actually, the history of modern rainmaking marked its first beginning in 1931 when Varaat of Holland dropped dry ice, among other things, into super-cooled clouds and produced slight amounts of rain. The possibility of producing rain from supercooled clouds by the introduction of artificial nuclei was also foreseen by the German cloud physicist Findeisen in 1938, but it was not until 1946 when a satisfactory method of supplying nuclei in the required quantity was discovered. The first time, Vincent J. Schaefer, an American meteorologist and science consultant succeeded in putting 6 pounds of crushed dry ice into a supercooled altocumulus cloud near Greylock Mt. in western Massachusetts on Nov. 13, 1946. In 1947, Vennegut discovered the effectiveness of silver-iodide as a cloud seeding material. This was followed the next year by Langmuir's demonstrations that plain water could sometimes trigger the precipitation process in warm clouds by developing a sort of chain reaction in droplet growth. Since then many efforts have been made by meteorologists to find a cloud-seeding material better than silver iodide or dry ice, they are still the only materials currently in use for large-scale operations directed towards modifying supercooled clouds.

Though, during this long span of time, a change in techniques of introducing freezing nuclei into clouds has been observed. First of all, freezing nuclei like dry ice or silver-iodide were delivered only by airplane into the clouds. Later a search for better freezing nuclei than dry ice or silver-iodide led to the disclosure that smoke of silver-iodide also acts as still a better freezing nuclei at temperatures as high as -60C. Then it was considered not to be necessary to apply an expensive technique of delivering silver iodide by airplane as usual, but it could be introduced in the form of smoke from the ground itself with the help of ground generators or surface burners. This idea was conceived from the Indian tradition of performing Yajñas. Actually, these ground generators are the modern representatives of the ancient Yajña kuṇḍas or fire altars.

The technique of seeding clouds with the help of surface burners is very cheap compare to that of an airplane or firing artillery shells, a very old idea revived by Russian scientists. The technique of Yajña-kuṇḍas or surface burners has only one drawback that is to rely on the air currents to carry the smoke up into the atmosphere. In this method, with no control over the subsequent transport of the smoke, it is not possible to make a reliable estimate of the smoke reaching into upper spheres. To overcome this problem, the smoke can be dispersed from burners in aircraft. This idea was also being practised in ancient India, as has been revealed in the Bṛhadvimānaśāstra, an ancient Indian treatise on the science of aeronautics. Airplanes were equipped with the Dhūmodgama or Dhūmaprasāraṇa instruments to disperse the smoke in the atmosphere so that the ill-designed bids of an enemy might be foiled, e.g. one verse of the Bṛhadvimānaśāstra (15.65) reveals how the airplane was used to be protected from enemies' attack by dispersing poisonous smoke from the Apasmāradhūmaprasāraṇa Yantra fitted in the plane.

svayāna rakṣaṇārthāya parayānair yathāvidhi
apasmāra dhūmaprasāraṇa yantraṁ pracakṣate.

In spite of the best thorough-going efforts of meteorologists in the field of rainmaking, it can be concluded that the modern science and technology of rainmaking is still in its infancy. It is confined only up to cloud modification. It has not so far been able to go beyond this. It doesn't see any prospects or scope for cloud formation. Moreover, the results of trials carried out in the direction of cloud seeding have not been so encouraging and impressive. To sum up, it can be said that modern scientists have just made a beginning in the direction of weather modification and still they are to go a long way up to give a concrete and scientific shape to their immature start. Perhaps, that is why, they have not been able even to speculate to modify the direction of winds, to form clouds, to bring seasonal changes and to ward off the falling rain. Thus even after a long consistent effort of study and research into the field of weather modification, the achievements of the modern science of weather modification may be calculated as nil.

Rainmaking by Vedic Scholars

In addition to the various efforts made in the direction of rainmaking as laid down above, one more attempt has been made since time immemorial and that is with the help of Yajña. This technique has a long-standing history and has reached us through an age-old long tradition. That tradition is still alive in the historical records of our Saṁhitās, Brāhmaṇas. Āraṇyakas, Upaniṣads, Sūtras, Epics and Purāṇas and other medieval and modern literature pertaining to the Vedas, Upavedas or Vedāṅgas. First of all, in the history of humankind, the idea of inducing rain with the help of Yajña was conceived by a Vedic seer, who in the very explicit terms proclaimed as vṛṣṭiśca me yajñena kalpatām.[2] 'Get me rains with the help of Yajña.' Perhaps, keeping in view the authenticity of this technique, another Ṛṣi proclaimed 'nikāme nikāme naḥ parjanyo varṣatu'.[3] 'Let the cloud precipitate as and when we desire.' This science of rainmaking didn't end with these proclamations of the Vedic Ṛṣis, but in course, rainmaking agents were also discovered and it was made known that Mitra and Varuṇa were the rainmaking agents.

mitrāvaruṇau tvā vṛṣṭyāvatām.[4]

'May the Mitra and Varuṇa bring rain for you.'

'mitrāvaruṇau vṛṣṭyādhipati tau māvatām.[5]

'May the rainmaking agents, Mitra and Varuṇa, protect you.'

The other Ṛṣis made further advances in the area and could come to the conclusion that the co-ordination of both the elements is necessary for inducing rain and the proposed coordination could easily be effected with the help of Yajña.

yajñā no mitrā varuṇā yajñā devam ṛtaṁ bṛhat.[6]

'Mitra and Varuṇa,' the main agents of rain, should be coordinated with the help of Yajña for the precipitation of rain

Thus the Yajña became the part and parcel, in fact, the basis of the Vedic life. Perhaps, that is why, the Vedic Ṛṣi didn't hesitate in passing the direction to every human being to perform Yajña season-wise daily, monthly and yearly to stimulate seasonal deities so that they may yield their favour for them as per their desires.

ṛtūn yajña ṛtupatinārtvānut hāyanān

samāḥ saṁvatsarān māsān bhūtasya pataye yajña.[7]

Following the notion of the Vedic Ṛṣi, the author of the Brāhmaṇas proclaimed Yajña as the noblest act.

yajño vai śreṣṭhatamaṁ karma.[8]

They also elucidated the process and role of Yajña in stimulating precipitation as:

agnervaidhūmo jāyate.
dhūmād abhram. abhrād vṛṣṭiḥ.[9]

'Yajñīya fire helps in smoke development, the smoke helps in cloud formation and clouds yield rain.'

The idea of rainmaking with the help of Yajña, which found its genesis in the Saṁhitās was developed in the Brāhmaṇas and Āraṇykyas and sought its culmination further in the Upaniṣadika literature. The Upaniṣadika Ṛṣi at one place doesn't want to miss the opportunity of disclosing the technique as to how to induce an existing cloud to rain. He speaks thus :

parjanyo vāgnir gotam tasya saṁvatsara eva samidabhrāṇi dhūmo vidyudarccirasanir aṅgārā hrādunayo visphuliṅgās-tasminnetasminn agnau devāḥ somaṁ rājānaṁ juhvati tasya āhutyai vṛṣṭi samabhavati.[10]

'That is to say briefly, an existing cloud may be stimulated to precipitate by the Soma āhuti released into the Yajña.'

The same idea was further inherited by the Epics and Purāṇas. Lord Krisna in his massage to the dejected and disappointed Arjuna while relating the overall dominance of Yajña, describes the fact that the Yajña is helpful in inducing rain. The verses go like this:

annād bhavanti bhūtāni

parjanyād anna sambhavaḥ.

yajñād bhavati parjanyo

yajñaḥ karma samudbhavaḥ.[11]

karma brahmodbhavaṁ viddhi,

brahmākṣara samudbhavam

tasmāt sarvagataṁ brahma nityam

yajñepratiṣṭhitam.[12]

In addition to this, the Bṛhadvimānaśāstra an ancient treatise on the science of aeronautics, treats this subject in a more technical and mechanised way. It gives a detailed break-up as to how to develop various types of scientific instruments to save the airplane from all those natural calamities which are likely to cause damage to it and its crew. For instance, it proposes on the basis of Yantrasarvasva to develop a Śiraḥkilaka instrument to dispel the effect of lightning.

yadapāyo vimānasya bhavedaśanipātataḥ

tadapāya nivṛttya-rthaṁ śiraḥkilaka yantrakam.[13]

Further, it lays down the techniques as to how to develop, among several other instruments, the Varṣopasaṁhāra, Tryāsyavātanirsana and Ātapopasaṁhāra instruments to neutralise the effects of rain, winds and heat respectively, which are likely to affect the aeroplane. e.g.

uktāni hi yantrasarvasve yantratrayaṁ yathā vidhi.[14]

sarveṣāṁ sukhabodhāya tānyevatra pracakṣate

tryāsyavāta nirasana yantraṁ tadvān manoharam.[15]

sūryātaposaṁhāra yantraṁ caiva tataḥparam

ativarṣopasaṁhāra yantraṁ ceti tridhā smṛtam.[16]

To protect the airplane from the opposition of strong air, the Vimānastambhan instrument was stated to have been installed in the airplane, e. g.

vātapravāhasaṁasarga parihārāya kevalam

vimānastambhanayantraṁ yathāmati nirupyate.[17]

Swami Dayananda's Revelation

In modern times, the age-old idea of rainmaking which fizzled out with the passage of time was revived by Swami Dayananda Saraswati, a great embodiment of Vedic life and thought. He made revolutionary observations on the various burning issues concerning political, social, economic, religious, and linguistic aspects of life and on the interpretation of ancient Indian language and literature. Among an immense number of things, he observed on the basis of the evidence from the Vedas that the rain could be induced with the help of Yajña as and when we desire. During the course of relating various purposes of Yajña, he gives a detailed description of the process of inducing rain with the help of Yajña. He bases his ideas on the evidence of Śatapatha Brāhmaṇa 5.3.5.17. According to him,

agneḥ sakāśād dhūmavāṣpau jāyete. yadā yamagnirvṛkṣauṣadhivanaspati jalādipadārthān praviśya tān saṁhatān vibhidya tebhyo rasaṁ ca pṛthak karoti, punaṣte laghutvamāpannā vāyvādhār-eṇoparyyā kāśaṁ gacchanti. tatra yāvān jalarasā-ṁśastāvato vāṣpasaṁjñāsti. Yaśca niḥsneho bhāgaḥ sa pṛthivyaṁśo'sti. ata evobhaya bhāgayukto dhūma ityupacaryate. Punar dhūmagaman-ānantaramākāśe jalasaṁayo bhavati. tasmādabharaṁ ghanā jāyante. tebhyo vāyudalebhyo vṛṣṭirjāyte.[18]

'That is, due to the treatment with fire, carbon and vapour are generated. When the fire enters the trees, herbs, plants and water, etc. it decomposes them and separates their juicy substance from them, which on account of its lightness, is accelerated upward towards the sky in conjunction with air. In this process, the juicy substance which is an aqueous form is known as vapour and what remains dry is known as carbon which is the part of the earth. So both forms earthen and aqueous in a combined state are known as smoke. After the smoke reaches the sky, water vapours accumulate there. These vapours give birth to clouds and clouds yield rain.'

Rainmaking by Ram Narain Arya

Following Dayananda's revolutionary observations, the idea of rainmaking with the help of Yajña gained a momentum and found new support at the hands of Āryasamājī scholars. A few of them even came forward who endeavoured to induce rain or to prevent it, but their experiments couldn't yield fruitful results.[19] Here in this regard, a particular mention may be made of Ram Narain Arya, M.A. (1936-2010) who can be called the perfect rainmaker of the twentieth century. After spending 35 long years of study, research and experimentation, he had come to a startling conclusion what he called the Vedic way to beat nature. Not only had he been able to induce rain successfully, but he had also been forming clouds, changing the flow of air in terms of direction and speed, stopping the falling rain and even modifying into the weather conditions, from dry to wet, hot to cold and vice-versa for the last 36 years. The author of the present lines has been a close associate and witness to most of the rainmaker's experiments carried out by him from time to time at various places in India. Having been confirmed strongly of the authenticity and success of the rainmaker's experiments, this author thought it viable to bring the rainmaker's research into the limelight so that not only the people of India but the people of the whole world may be benefitted from the experiences and research of the rainmaker. Actually, the rainmaker has preserved his ideas and experiences, with regard to his experiments on weather modification like rainmaking, prevention of rain, modification into the direction and speed of airflow, modification into the other weather conditions, prevention of pollution and diseases in his daily diaries written by him from time to time. This author could get the privilege to make good use of the subject matter enshrined in the daily diaries (DD) of the rainmaker. At the places of apprehension and doubts, he made serious discussions with the rainmaker to clarify the same. To proceed further, it is necessary to render hereunder the brief life sketch and work of the rainmaker.

A Brief Life-sketch of the Rainmaker

Sh. Ram Narain Arya was born on Amāvasyā (the new moon day) of the Srāvaṇa month in 1993 of the Vikram era (or 14th July 1936) in the village of Farmana, District Sonepat, Haryana. His mother breathed her last when he was in his infancy. Under the circumstances, he along with his two brothers and one sister was brought up by his father. His father, Sh. Ratti Ram Arya was fond of physical exercises and wrestling. A devout and religious fellow, he was a devotee of the Veda. Ram Narain was greatly influenced by his father and inherited his qualities of doing Yoga and physical exercises regularly. He completed his schooling at the village high school. Afterwards, he passed out the examination of I.G.D. Bombay from Baraeli and was appointed in 1955 as a Drawing teacher in the Govt. Higher Secondary School, Bahujholari (District Jhajjar of Haryana). Late in the forties, he completed his B.A. and did his M.A. in Political Science and Sanskrit (Veda). From childhood, he was very intelligent, curious and studious. He had the chance to study the literature of Kabeer, Raheem, Tulsidas, Raidas and Guru Nanak. His wife, Prem Vati Prabhakar, who is also a religious lady, used to read him the stories of Rāmāyaṇa, Mahābhārata and Gitā. All this added to his detachment from the worldly allurements. Gṛhastha as he is, he is leading a life of an ascetic. In 1955, he came in contact with Bhagvandeva Acharya (Swami Omananda), Acharya Baladeva and Brahmachari Indradeva (Swami Indravesh) at Gurukul Jhajjar. This new acquaintance promulgated him to study the Vedas. He then moved to the study of the Vedas, Upavedas and Vedāṅgas. Influenced by Swami Dayananda's works such as Satyārhta Prakāsh and Samskāravidhi, he started performing Yajñas daily at the time of sunset and sunrise. During his studies of the Vedas and Dayananda's works, he came across such references as could give him an idea of rainmaking with the help of Yajña. He also made an in-depth study of the works of Āryabhaṭṭa, Varāmihira and Bhāskarācārya which added to his knowledge of geology and astronomy. He had a chance to make an intensive study of the Bṛhadvimānaśāstra, an ancient Indian treatise on the science of aeronautics composed by Maharishi Bharadvāja. This work closeted him with the different aspects of ancient Indian science related to weather modification. Thus having taken cues of rainmaking with the help of Yajña from the Vedas, Upavedas, Vedāṅgas and Dayananda's works and having acquired the knowledge of the properties of the matter as described therein, he all set to perform Yajñas in a particular direction. It was to modify weather, modify airflow in terms of direction and speed, to induce rain and to prevent the same, to remove famines, to prevent deluges, diseases and to beat the pollution.

The experimental Yajñas performed strictly in conformity or in tune with the Vedic principles and surprisingly enough everything was witnessed taking place actually which was considered sometimes ago a fallacy or mythology. He calls this method of controlling natural powers with the help of Yajña as a Vedic way to beat nature. Not only did he do experiments with the nature outside, but he also practised Yoga to attain self-accomplishment and applied it to detect the deficiencies or inconsistencies in the matter of natural course. According to him, Yajña and Yoga should go side by side. By way of Yoga, one can feel the nerve of nature and detect its deficiencies or inconsistencies like an able and dexterous physician does with a patient and by way of Yajña, medication of nature can be done to remedy all its deficiencies and inconsistencies. His dedication to this cause was so great that all the institutes he served were converted by him into laboratories for his experiments on nature. He served at the following institutions.

G.H.S.S. Bahujholari (1955-66); G.M.S. Ruraki (1967-68); G.M.S. Guhna (1-4-1968 to 24-9-1968); G.H.S. Khanpur Kalan (25-9-68 to 7-10-68); G.H.S. Mundalana (11-10-1968 to 18-7-1973); G.M.S. Rithal (19-7-1973 to 31-7-1974); G.G.H.S. Bhainswal Kalan (1-8-1974 to 18-5-1976); G.M.S. Katwal (19-5-1976 to 1-8-1978); G.H.S. Bhainswal Kalan (2-8-1978 to 4-7-1979); G.M.S. Jasrana (5-7-1979 to 7-9-1989); G.H.S. Anwali (8-8-89 to 8-12-90); G.H.S. Bichpari (9-12-90 to 21-7-91). All these schools have been the main centre of his experiments.

Having been confirmed of the validity and authenticity of the Vedic principles, he became a staunch exponent of the Vedic life and thought. He made it his life mission to propagate Vedic teachings and thought among the general masses, particularly youths and for that matter, he used to deliver prayer time discourses before students on the Vedas and their ancillary sciences during his teaching days. He, thus, made them known about the glorious past of India and ancient Indian advancement in the field of philosophy, sociology, polity, theology, science and technology. He voluntarily retired from the Govt. service in 1991 in order to speed up his activities for the furtherance of this noble cause. After retirement, he speeded up his mission of spreading scientific knowledge of the Vedas. He toured extensively and delivered more than 1000 public and popular lectures in different educational institutions and Jails to enlighten the students and jailed persons of the rich scientific heritage of ancient India and exhorted them to go for Yoga and Yajña for better health and healthy environment in society. He warns the younger generation not to fall a prey to the cultural pollution that is posing a great danger to humanity at large. He himself leads a very simple and austere life. His whole life has been a life of relinquishment and penance. Through Yoga and penances, he has so regulated his life that he spends the chilly winters away in simple summer wears. He takes simple sāttvika meals often saltless, chilly-less and sugarless. According to him, prior to the regulation of nature, regulation of the self is a must.

Rainmaking with the help of 'Yajña'

As already indicated above, modern meteorologists are seeking the possibility of producing rain from super-cooled clouds by introducing some freezing nuclei therein. In their concerted efforts, they have been able to discover solid carbon dioxide (dry ice) and silver iodide as successive better cloud seeding materials. Their further efforts could not reveal any improvement, rather could bring a change in techniques of introducing freezing nuclei in clouds. First of all, freezing nuclei like dry ice or silver iodide were delivered only by airplane into the clouds. Later a search for better freezing nuclei than dry ice or silver iodide led to the disclosure that the smoke of silver-iodide also acts as still a better freezing nuclei at temperatures as high as 600 C. Then it was considered not to be necessary to apply the expensive technique of delivering silver iodide by airplane as usual, but it could be introduced in the form of smoke from the ground itself with the help of ground generators or surface burners.1 This idea was conceived from the Indian tradition of performing Yajñas. Actually, these ground generators are the modern representatives of the ancient Yajña Kuṇḍas or Fire-altars. The technique of seeding clouds with the help of surface burners is very cheap compared to that of an airplane or firing artillery shells, a very old idea revived by Russian Scientists, the technique of Yajña-Kuṇḍas or surface burners has only one drawback that is to rely on the air currents to carry the smoke up into the atmosphere. In this method, with no control over the subsequent transport of the smoke, it is not possible to make a reliable estimate of the smoke reaching into upper spheres. To overcome this problem the smoke was dispersed from surface burners in aircraft. This idea was also practised in ancient India, as has been revealed in the Bṛhadvimānaśāstra, an ancient Indian treatise on the science of aeronautics. Airplanes were equipped with the Dhūmodgama or Dhūmaprasāraṇa instruments to disperse the smoke in the atmosphere.

In spite of the best thoroughgoing efforts of meteorologists in the field of rainmaking, it can be concluded that the modern science and technology of rainmaking is still in its infancy. It is confined only up to cloud modification it has not so far been able to go beyond this. It doesn't see any prospects or scope for cloud formation. Moreover, the results of trials carried out in the direction of cloud seeding have not been so encouraging and impressive. To sum up, it can be said that modern scientists have just made a beginning in the direction of weather modification and still they are to go a long way up to give a concrete and scientific shape to their immature start. Perhaps, this is why, they have not been able even to speculate to modify the direction of winds, to form clouds, to bring seasonal changes and to ward off the falling rain. Thus, even after a long consistent effort of study and research into the field of weather modification, the achievements of the modern science of weather modification may be calculated as nil.

In addition to the various efforts made in the direction of rainmaking as laid down above, one more attempt has been made since the time immemorial and that is with the help of Yajña (Here it may be noted that there are many types of Yajñas such as Śrauta yajñas, Gṛhya Yajñas and Pañca-mahā Yajñas like Pitṛ yajñas, Brahma Yajña, Bali-vaiśvadeva Yajña, Atithi-Yajña and Deva-Yajña. In the present context, Yajña means Deva Yajña). This technique has a long-standing history and has reached us through an age-old long tradition. That tradition is still alive in the historical records of our Saṁhitās, Brāhmaṇas, Āraṇyakas, Upaniṣads, Sūtras, Epics and Purāṇas and other mediaeval and modern literature pertaining to the Vedas, Upavedas or Vedāṅgas. First of all, in the history of humankind, the idea of inducing rain with the help of Yajña was conceived by a Vedic seer, who in the very explicit terms proclaimed as

vṛṣṭiśca me yajñena kalpatāṁ[2]

'Get me rains with the help of Yajña.'

Perhaps, keeping in view, the authenticity of this technique, another Ṛṣi proclaimed

nikāme nikāme naḥ parjanyo varṣatu[3].

'Let the cloud precipitate as and when we desire.'

This science of rainmaking didn't end with these proclamations of the Vedic Ṛṣis, but in course, rainmaking agents were also discovered and it was made known that Mitra and Varuṇa were the rainmaking agents.

Mitrāvaruṇau tvā vṛṣṭyāvatāṁ[4].

'May the Mitra and Varuṇa bring rain for you.

Mitrāvaruṇau vṛṣṭyādhipati tau māvatām[5].

'May the rainmaking agents, Mitra and Varuṇa, protect you.

The other Ṛṣis made further advances in the area and could come to the conclusion that the co-ordination of both the elements is necessary for inducing rain and the proposed coordination could easily be effected with the help of Yajña.

yajñā no mitrāvaruṇā yajñā devaṁ ṛta bṛhat.

'Mitra and Varuṇa, the main agents of rain should be co-ordinated with the help of Yajña for the precipitation of rain.'

Thus the Yajña became the part and parcel, in fact, the basis of the Vedic life. Perhaps, this is why, the Vedic Ṛṣi didn't hesitate in passing the direction to every human being to perform Yajña season-wise, daily, monthly and yearly to stimulate seasonal deities so that they may yield their favour on them as per their desires.

Ṛtun Yajña ṛtupatinārtvānut havyānāṁ samāḥ saṁvatsarān māsān bhūtasya pataye Yajña.[7]

Following the notion of the Vedic Ṛsi, the Brāhmaṇkāras proclaimed Yajña as the noblest act.

Yajno vai śreṣṭhatamam karma.[8]

They also elucidated the process and role of Yajña in stimulating precipitation as:

agnervai dhūmo jāyate. dhumād abhram. abhrād vṛṣṭiḥ.[9]

'Yajña-fire helps in smoke development, the smoke helps in cloud formation and clouds yield rain.'

The idea of rainmaking with the help of Yajña, which found its genesis in the Saṁhitas developed further in the Brāhmaṇas and Āraṇyakas and sought its culmination in the Upaniṣadic literature. The Upaniṣadic Ṛṣi at one place doesn't want to miss the opportunity of disclosing the technique as to how to induce an existing cloud to rain. He speaks thus:

Parjanyo vāgnirgotam tasya saṁatsara eva samidābhraṇi dhumo vidyud arccirasanir aṅgārā hrādunayo visfuliṅgās tasminn etasminn agnau devāḥ somaṁ rājānaṁ juhvati tasya āhutyai vṛṣṭi saṁbhavati.[10]

> 'That is to say, O' Gotam! Parjanya (the cloud) is the second sacrificial altar. It has year as its samidhā (sacrificial fuel), abhra (supercooled cloud) as its smoke, and lightning as its flame, aśani as its ember and the thunder-sound as its spark. This cloud can be induced to rain by way of augmenting the somīya element in the atmosphere with the help of somīya āhutis.'

The same idea was further inherited by the Epics and Purāṇas.

Further, it lays down the techniques as to how to develop, among several other instruments, the Varṣopasaṁhāra, Tryāsyavātanirsana and Ātaposaṁhāra instruments to neutralise the effects of rain, winds and heat respectively, which are likely to affect the airplane, e.g.

uktāni hi yantrasarvasve yantratrayaṁ yathāvidhi.[14]

sarveṣām sukhabodhāya tānyevātra pracakṣate

tryāsyavātanirasanayantram tadvān manoharam.[15]

sūryātaposaṁhārayantraṁ caiva tataḥparam

ativarṣopasaṁhārayantraṁ ceti tridhā smṛtam.[16]

To protect the airplane from the opposition of strong air, Vimānastambhana instrument was installed in the airplane, e.g.

vātapravahasaṁsarga parihāraya kevalam.
vimānastambhanayantra yathāmati nirupyate [17]

Method of Rainmaking with the help of 'Yajña'

In fact, rainmaking with the help of Yajña as stated earlier is a systematic process of co-ordinating the rainmaking agents, Mitra and Varuṇa. For instance, the *Vājasaneyī Saṁhitā* (VS.) Says:

Yajñā no mitrāvaruṇā yajñā devaṁ ṛtaṁ bṛhat.[19]

Mitra is a soma element or āpastattva, or say negative charge in modern scientific terms, Varuṇa is an āgneya element or jyotiṣtattva, i.e. a positive charge. Co-ordination of both the elements in a particular ratio stimulates rain.

Apāṁ ca jyotiṣaśca miṣri bhāvakarmaṇo varṣakarma jāyate.[20]

The deficiency of any of these elements in the atmosphere may be covered up with the help of their respective āhutis offered in the terrestrial fire or Yajña-fire. Actually, Yajña acts as the remote controller and coordinator for the various deities, i.e. the natural elements existing in the mid-sphere or celestial sphere. Rainmaking with the help of Yajña is a function of maitrāvaruṇi āhutis offered to the fire in a particular ratio subject to the position of the two elements in the atmosphere. So, a rainmaker has a tough task before him. He will have to be very vigilant and constantly keep himself in touch with the weather conditions prevailing in the atmosphere. For rainmaking, he doesn't have to be dependant, as the other meteorologists have to be, on the pre-existence of the clouds. In fact, he is to start from zero and modify weather, as the need be, to any such an extent as to blow moisture-bearing winds, to generate evaporation, to seed precipitation yielding clouds, or to create other similar conditions leading to rain-fall or simply to induce rain from the already existing clouds. This fact is also very well supported by the Brāhmaṇakāras following observations regarding the precipitation of rain with the help of Yajña.

agnervaidhūmojāyate.dhūmad abhraṁ abhrāt vṛṣṭiḥ.[21]

That is to say, when maitra āhutis or somīya āhutis are offered to the Yajñīya fire, the smoke (the vikāra of waters existing in the mid-sphere) is produced, which, due to the law of specific elementary gravity (See for Details: Vedic Meteorology), goes higher up in the atmosphere (*divaṁ te dhūmo gacchatu*)[22] through the fire column established with the help of celestial fire, the Sun, in conjunction with the air, (*marutaṁ pṛṣati gaccha vasā pṛṣnirbhūtvā divaṁ gaccha*)[23]. There it creates abhra (clouds) and clouds precipitate rain (tato no vṛṣṭi-imāvaha).[24]

Thus the rainmaker, with the help of Yajña, is capable of seeding clouds and stimulate them to rain as and when desired. But in the process, before starting a Yajña at any place or time, it is a priory of a rainmaker to make a proper survey of weather or study weather conditions properly. The weather is to be studied in terms of air-motion, i.e., whether the air draft is in a favourable direction or not; in terms of humidity or water vapours, i.e., whether the air is humid enough to reach saturation point or dry requiring further evaporation of water from the ground, and in terms of cloudiness, i.e., whether the sky is seeded with clouds or there is no cloud at all. Thus a serious rainmaker should know the weather conditions properly and exactly before carrying out a rainmaking experiment. Different weather conditions would oblige a rainmaker to modify weather differently for rainmaking and this would require different time spans and varied expenses to induce rain with the help of Yajña.

Thus from the foregoing, it is crystal clear that the induction of rain or rainmaking depends exclusively on the current weather conditions. The weather may be dry cum hot; dry cum cold; humid cum hot; humid cum cold and accordingly air may be tropical (warm), polar (cold), maritime (humid), or continental (dry). With the help of Yajña (Deva yajña), one can modify weather to the extent that it shall favour rainmaking. Following modifications of weather may be effected with the help of Yajña in order to induce rain.

Evaporation and cloud formation

If the weather is dry and hot and the west-wind (continental air) is under sway, a rainmaker is required to do *vāyuviloḍana* or *ākāśa-manthana*, i.e., an atmospheric stirring. For this, the air is heated up and made to flow around turbulently so that a high evaporation of water from the ground or sea may easily be generated. See for example the following hemistich of the RV.

Saṁ no iṣiro vātu vātaḥ.[25]

'let the rain-precipitating winds blow from all

directions.'

This is well-supported by the following verse of the AV.

Prajāpati salilādā samudrādāpa

Iryannudadhimardayāti.

Udīryat marutaḥ samudratastaveṣo

Arko nābhā ut pātyātha.[26]

'The Sun assisted by air evaporates the terrestrial waters into celestial ones and makes them fall on the earth in the form of rain.'

A similar observation is made by the Bṛhaddevatā.

Rasāna raśmibhirādāya,

Vāyunāyaṁ gataḥ saha

Varṣatyeṣa ca yalloke

Tenendra iti smṛtaḥ.[27]

'The Sun in conjunction with air conducts evaporation of fluid substances from the earth.'

For this type of heating and airflow Vāruṇi āhutis are offered to the Yajña. It augments the power of the Varuṇa deity or the *āgneya* element in the atmosphere. Thus strengthened by *āgneya ahutis*, the varuṇa element heats up the atmosphere and makes the air flow strongly. Thus two-three days of atmospheric stirring enables air to receive sufficient moisture required for saturation. Afterwards, condensation starts and clouds begin to form. But this is not sufficient to induce a good amount of rain. This is only the first phase of the process.

Persistence of sea bearing winds

For sufficient rain, it is necessary for the moisture/rain/sea bearing winds and clouds to persist. In the context of the Indian sub-continent, as it is well known, only the east winds/monsoons bring moisture and rains.

Hence, with the help of Yajña the somīya substance is sprayed in the atmosphere to augment the somīya content of the air, which helps deflect the sea-bearing winds.

Persistence of clouds and precipitation of rain

For a cloud to persist, Yajña-kuṇḍa is maintained continuously day and night in order to keep the air ascending/slightly moving upward. Actually, when the Yajña is performed, the surrounding air is heated up and gets an upward thrust. This fact is also corroborated by the Vedic seer as:

swāhākṛte   ūrdhavaṁ   nabhasaṁ   marutaṁ gacchatam.[28]

'When the Yajña is started, the air is accelerated upward.'

This upward moving air helps the clouds induce rain. This function of air is confirmed by the following authorities of Vāyupurāṇa.

meghā vāyunighātena viṣṛjanti jalaṁ bhuvi.[29]

abhrasthāḥ prapatantyāpo vāyunā samudiritāḥ.[30]

Maitrāiṇi Saṁhitā also bears out the same fact, as;

maruto'mutaścyāvayanti etc vai vṛṣṭyāḥ pradātāraḥ.[31]

This is also well-supported by the following authorities of the AV. e.g.,

marudbhiḥ prachyutā meghā varṣantu pṛthivimanu.[32]

marudbhiḥ pracyutā meghā samyantu pṛthivimanu [33]

samabhrāṇi vātajūtāni yantu.[34]

upapruto marustanairyata vṛṣṭiryā viśvānivatasprṇāti.[35]

This process, as the experience had it, will take three to fifteen days to induce rain.

Secondly, there is likely that the favourable wind, i.e., sea bearing wind is already blowing and the sky is also cloudy. Under such conditions, a rainmaker is required to introduce such somīya contents in the atmosphere as could act as super-seeding agents. Clouds thus superseded would be able to yield rain within 24 or some more hours. This fact has been disclosed by the Bṛhadāraṇyakopaniṣad thus,

parjanyo vāgnirgotama tasya samvatsara eva samidabhrāṇi dhumo vidyudarcciraśanir aṅgārā hrādunayo visfuliṅgāstasminnetasminnagnau devaḥ somam rājānam juhvati tasya āhutyai vṛṣṭiḥ sambhavati.

'O Gotam! parjanya (the cloud) is the second altar. It has the year as its samidhā (sacrificial fuel), abhra (supercooled cloud) as its smoke, lightning as its flame, aśani as its ember and the thunder sound as its spark. This could be induced to rain by way of augmenting the somīya element in the atmosphere with the help of somīya āhutis.'

Nature of Āhutis

The aforementioned discussion regarding the method of inducing rain with the help of Yajña remains uncomplete so long as the nature of somīya and vāruṇī āhutis is not defined. Hence, a detailed break-up of somīya and vāruṇī āhutis in respect of various climatic conditions is rendered hereunder. To start with, it may be pointed out that for the sake of rainmaking or anti-rain experiments, climatic conditions should be studied into six types in the light of the variables of coldness and hotness in respect of humidity and aridity.

In this way the six climatic types to be worked out will be as:

1. Maritime polar (humid cum cold),

2. Maritime tropical (humid cum hot),

3. Maritime temperate (humid cum equable).

4. Continent polar (arid cum hot),

5. Continental tropical (arid cum hot) and

6. Continental temperate (arid cum equable).

Similarly, *āhuti dravya* (offering material) is composed of *auṣadhis* (vegetation) grown in various climatic conditions. From the point of growth of vegetation to be used as *āhutis* in humid or arid regions, the vegetations are classified as somīya and āgneya respectively and so are the *āhutis*.

Somīya Āhutis

As is clear from the foregoing discussion, somīya āhutis are composed of somīya flora or vegetation growing in humid regions. They may be classified into three categories in respect of their growth in high, low, or equable temperatures. For instance :

(1). Somīya or maritime flora growing in polar

'*varṣavṛddhamasi prati tvā varṣavṛddaṁ vattu.*'[33]

'Since you have grown in rains, one should know you as rain promotor.'

While commenting on this verse, Ś.Br. exemplifies the rain-promoting agents like rice, reed and bamboo. For example, it had it as :

'*atha śūrpam ādattle varṣavṛdhaṁ hyetadyadi naḍānāṁ yadi vetūnām yadīṣikāṇāṁ varṣamuhyevaitā vardhayati.*'[34]

'He then, in the application of first part takes hold of a winnowing basket made of dried straw, reeds, or bamboo. Since, straw, reeds, or bamboo have grown in rain, promote rain'

Further in connection with the second part of the verse i.e.

prati tvā varṣavṛddhaṁ vettu, he maintains as 'atha havirnirvapati, prati tvā varṣavṛdhaṁ vettviti. varṣavṛddhā uhyevaiti. yadi vrīhiyo yadi yavā varṣamuhya ivaitān vardhayati tatsaṁjñām eva etat śūrpāya ca vadati nedanyo'nyaṁ hinsāta iti'.[35]

'In application of the second part, the oblation is prepared to be offered to the Yajñiya fire in the winnowing basket. The oblation is made of rain-promoting vegetation. Rice and barley grow in rain and so promotes rain. Hence they are taken for oblation in the winnowing basket made of dried straw, reeds, or bamboo. Rice and barley are collected in the basket of straws, reeds or bamboo since both rice and straw/reeds/bamboo are rain promotors. They are not anti to each other.'

Thus in view of the facts and circumstances discussed above, it can unhesitatingly be maintained that rainmaking and prevention of rain is a matter of co-ordination between the six climatic conditions and six types of vegetations to be offered in the Yajñiya fire depending upon the careful selection and handling of āhuti dravya (vegetation) in respect of the prevailing climatic condition.

Eventually, it can be held that this thrilling idea of rainmaking or anti-rain with the help of Yajña is not only cherished by the present author and the Rainmaker, but modern scientists have also reached to some nonetheless thrilling conclusions, such as (1) the better effectiveness of silver-iodide as a cloud seeding material when introduced in the sky in the form of smoke from the ground with the help of surface burners or ground generators than when delivered by airplane. (2) Burning sugarcane in eastern Australia is thought to have reduced precipitation near the burning areas through over-seeding.[36]

These conclusions of modern meteorologists also validate the authenticity of the idea of rainmaking and anti-rain with the help of Yajña (Deva yajña).

To sum up, it can unhesitatingly and safely be said that the Vedic science of Yajña is based on concrete and scientific ideas which need to be revived and revitalised. Through these lines, the author would like to invite the attention of the World Governments and scientific bodies that they should come forward to save this scientific treasure of the Vedas from being threatened by extinction and preserve, popularise and promote this ancient tradition of Yajña by establishing Yajña saṁsthānas.

Notes & References

1. Mason (1975: 124)

2. *VS.* 18.9

3. *Ibid.* 22.22

4. *Ibid.* 2.16

5. *AV.* 5.24.5.

6. *VS.* 33.3

7. *AV.* 3.10.9.

8. *Kath.S.* 30.10; *Kath.S.* 46.8; *TS.* 3.2.1.4; *S. Br.* 1.7.1.5.

9. *S.Br.* 5.3.5.17.

10. *Bṛhadāraṇyakopaniṣad,* 6.2.10.

11. *Gita,* 3.14

12. *Ibid.* 3.15

13. *Bṛhadvimānaśāstra,* 13.19.

14. *Ibid.* 22.48.

15. *Ibid.* 22.49.

16. *Ibid.* 22.50.

17. *Ibid.* 15.91.

18. For details, see the author (1993:44 - 66), also (1995)

19. *VS.* 33.3.

20. *Nir.* 2.16.

21. *S.Br.* 5.3.5.17.

22. *VS.* 6.29.

23. *VS.* 2.16.

24. *Ibid.*

25. *RV.* 1.35.4.

26. 4.12.2.

27. 1.68.

28. *VS.* 6.11.

29. *Vāyupurāṇa,* 51.15.

30. *Ibid.* 51.25.

31. 2.4.8.

32. *AV.* 4.12.8.

33. *Ibid.* 4.12.8.

34. *Ibid.* 4.12.1.

35. *Ibid.* 6.22.3.

36. Herbert (1965: 115)

References

Bṛhadvimānaśāstra (1959): by Maharshi Bhardwāja, Ed. with Hindi Translation by Brahma Muni Parivrajaka, Hardwar.

Hason, B.J. 1975: *Clouds, Rain and Rainmaking*, 2nd Ed. Cambridge University Press, 1975.

Herbert Riehl, 1965: *Introduction to the atmosphere*, 3rd Ed., McGraw-Hill Kogakusha Limited, 1965.

Kāṭhak Saṁhitā (KS): ed. by Sripāda Dāmodara Sātavalekara, Svadhyaya Mandal, Pardi.

Nirukta of Yāska (Nir.) (1959): by Brahma Muni Privrajaka, Ajmer Ravi Prakash Arya, 2006: Vedic Meteorology Indian Foundation for Vedic Science, Delhi.

Srimad Bhagvad Gita - A Vedic Scientific Scripture of Liberation (2015): by Dr. Ravi Prakash Arya, Amazon Books, USA.

Śatapatha Brāhmaṇa (Ś.Br.) with Sayana Bhashya and Commentary of Hariswami, published by Rashtriya Sanskrit Sansthan, Delhi.

The Vājasaneyī Saṁhitā (VS)(1997): ed. by Dr. Ravi Prakash Arya, Parimal Publication, Delhi.

The Taittirīya Saṁhitā (TS): ed by Sripāda Dāmodara Sātavalekara, Svadhyaya Mandal, Pardi.

The Ṛgveda Saṁhitā (RV) (1999). Ed. Ravi Prakash Arya, Parimal Publications, Delhi.

The Atharva Veda (AV.) (1982) of Śaunaka recension by Devi Chand, MA, Munshi Ram Manohar Lal, Delhi.

The Vāyupurāṇa with Hindi Translation by Ram Pratap Tripathi, Hindi Sahitya Sammelan Prayag.